I0134269

EQUIPMENT
Gulf War Poems

EQUIPMENT
Gulf War Poems

Anthony Aiello

Two Birds
BROOKLYN

Published by Two Birds
18 Herkimer Street
Brooklyn, NY 11216
www.twobirdspublishing.com

Copyright 2014 by Anthony Aiello

All rights reserved, including the right of reproduction in whole or in part in
any form.

Grateful acknowledgement to the other places these poems have appeared, at
times in slightly altered form.

Chicagopoetry.com: "M-16 Combat Assault Rifle"
Echo: A Polyglot and Cross Cultural Journal (echopolyglot.com), February
2004: "Ammo Dump," "Departure," "Homo Furens"
Echo: A Polyglot and Cross Cultural Journal (echopolyglot.com), June 2005:
"The Five Things: Fire—Direction—Control," "The Five Things: A View
of the Road," "The Five Things: Floodwaters," "The Five Things: Assault
Dimensions," "The Five Things: February Forecast"
New England Journal of Public Policy, Winter 2005: "Quit Paradise," "Setting
Out"
Poets Against the War (www.poetsagainstthewar.org), selected poems of the
week, 25 September 2003: "A Prayer," "Where's Joe?"
Red River Review (www.redriverreview.com), February 2006: "Departure,"
"Homo Furens"

Library of Congress Control Number (LCCN): 2014944443
ISBN-13: 978 0 9910042 0 1
ISBN-10: 0991004205

Cover Design: Tessa Wright

First Edition

for Monika

Contents

Setting Out

In an ideal war, nobody expected to see much of each other or the enemy. We were an area weapon; our maximum range was thirty kilometers, our minimum eight, meaning we had to be further than five miles from our targets, which could be no further than nineteen miles from us. An army pamphlet from 1985 tells us that the Multiple Launch Rocket System "is a highly mobile, rapid-fire, free-flight rocket system that complements cannon artillery in the counterfire and air defense suppression roles. It also supplements other fire support systems by engaging high-density mechanized targets during surge periods and provides interdiction fires against follow-on forces. Targets include troops, light equipment, target acquisition systems, logistic complexes, and command, control, and communications systems."

It turned out they were wrong about targets. One launcher could kill most anything in a square kilometer with its twelve rockets, each rocket aimed separately by the computerized fire control system. In the shit, we wasted everything that came at us and much that was running away, and plenty of Iraqis who just sat there with no idea we'd sent up rockets. While talking to a bunch of tankers from the 24th Mechanized Infantry Division, who we fought with in Iraq,

i

they told me, Birdie, and Jeff Tebbe that every time their tanks were sent into a new area they called in fire missions, and every time they got there, everything was dead, so all they needed to do was finalize the details. They thanked us for making the war easy.

Attention to detail is the discipline that grows from the tender and meticulous commitment to equipment—especially weapons—every GI learns from his first days in the army. In basic training, Drill Sergeant told us to love our M-16s like a woman. I didn't agree that an M-16 was anything like any woman I'd ever known, but I nonetheless lingered over its details, its shape and feel, its temper and kick, its sound and fury the way I'd learned to linger over a woman's face, legs, arms, back, belly, breasts, and the rest. These lessons served me well when later, in advanced individual training, I learned not a single weapon but a weapon system that consists of two major components: ammunition supply and rocketfire. As our instructors told us, learning both together allows you to load your cake and shoot it too.

The HEMTT, pronounced HE-mit, is the heavy expanded-mobility tactical truck: a thirty-three feet long, fixed-bed, forward-cab truck with a rear-mounted, folding crane and eight five feet tall tractor tires. The HEMTT was a monster truck of the most amazing kind. You could edge the huge bastards through the thickest Carolina pine woods, simply running over the smaller trees while threading your way slowly between the larger ones, though you sometimes ran those over too. They could drive through six feet of water, or haul ass at seventy MPH down the twisting hill roads at Fort Bragg, where I was stationed. In the desert, the trucks were almost impossible to get stuck, though we did plenty of that early on. The HEMTT's crane was used to sling Launch Pod Containers, which we called six packs. Each held six rockets in individual tubes mounted in an aluminum frame close to thirteen feet long, four wide, two and a half high. We could sling six packs as fast as the cranes allowed, spinning the tons by touch into place and aligning them rank by rank, when building an ammo dump, dress right dress and all that. Day or night, when resupplying launchers, our two-man crews could drop two pods here and two others over there, in five minutes or less, spaced just right so two launchers could load at the same time.

The HEMTT is one half of MLRS, the other half is the Self-Propelled Launcher-Loader, the SPLL, the spill: a tracked, light-armored vehicle twenty-three feet long, eight and a half high, and ten feet wide. Like the ammo guys, launcher crews knew their SPLLs as well as any of us knew our M-16s, which of course we all could strip, break down the bolt, clean everything, replace the bolt and reassemble the rifle in our sleep in under a minute. At a party, Drill Sergeant told us, the HEMTTs supplied the six packs and the SPLLs brought the pain.

And we brought it all right. We killed light equipment: trucks and jeeps and cars didn't stand a chance. We destroyed most bunkers and the few buildings we fired on. We also killed medium armor like BMP personnel carriers. And we killed tanks, even T-72s. We weren't supposed to kill tanks, especially not 72s, the best Soviet tank Iraq could buy. In tests our rockets didn't kill tanks. But we didn't test massed battery fire, all nine launchers firing twelve-rocket salvos as fast as they could shoot their loads. "MLRS operations are characterized by rapid emplacement, engagement, and displacement (shoot-and-scoot tactics) of widely dispersed launchers."

Expecting to fight the Soviets—who greatly outnumbered us, had excellent equipment and training, whose counterbattery fire and suppression resources were as good as ours—in the forests of Germany, we trained to fight a lonely war. "An MLRS battalion is assigned to each corps and a separate MLRS battery is organic to each armored and mechanized infantry division as part of the 203MM[a great big-ass cannon]/MLRS battalion. An MLRS battery is organic to the high-technology, light infantry division. The battery is in light artillery and rocket battalion. All MLRS firing batteries are organized identically and can operate independently for short periods." In practice, each launcher operated solo, supplied by two HEMTTs, whose crews left six-packs at resupply points the SPLLs would visit when needed.

The Iraqis were not the Russians. Our launchers fired as close to each other as safety allowed, and as much, and as fast as they could, with dozens of rockets pouring thousands of submunitions into target zones much smaller than a square klik. This wasn't the rainstorm of grenade-sized bomblets a single launcher produced. This was the Deluge without hope of rescue.

Not even in childhood action-hero fantasies did I imagine I would one day become a rainmaker. The day I joined my mother looked at me and said, you son of a bitch. Every adult I knew growing up had experienced life in war, and she couldn't believe I'd volunteered for the pain forced on her generation and her parents' too. Mom was born in Puerto Rico but grew up in Chicago, where I was born. She came of age during the Vietnam War and was active in protests and marches against the war and for civil rights, during which she became acquainted with Freddy Hampton before the FBI and Chicago police assassinated him. She married my father at fifteen, divorced him at eighteen, me two years old. My father, then twenty-one and addicted to heroin, had deserted from the navy three years before in 1970 when orders came down for his ship to deploy to Vietnam.

Mom hired a hippie lawyer named Jerry Brody, who worked pro bono for dodgers and deserters, and my father came away from his court martial with a general discharge that eventually some years later became an honorable. Imagine that.

Their good friend, Lad, my godfather was a crewman on the Honest John missile system in Germany after being drafted. In the neighborhood, no one believed it: his brother Norby was already in Vietnam and they thought the army couldn't draft all a mother's sons. We all know how the vets came back with little or no welcome. My Uncle Lad left for boot camp with no farewells. His friends didn't believe he was gone. Norby flew in helicopters over the deep green death in Vietnam, a Huey door gunner. Three times Norby's helo went down, shot to shit dropping troops. The last time, Norby and his badly injured crew chief were the only survivors of the crash. Norby carried his comrade through the jungle, but the chief was dead by the time they reached safety. Several years gone by, Norby died on his third try, by rope in the garage, after somehow surviving the multiple stabbings to his chest and slashed wrists of the second attempt, having failed with pills the first time. Uncle Lad found his brother's body still warm and brought him down. Then Lad went down, checked himself into a hospital for a while, came out a Jehovah's Witness named Jerome.

Mom's best friend, Laurie, married a vet named Pat, who liked to sit with the news on television, volume turned low, Beatles records playing loud (the *White Album* more often than not). Every day he sat there and read the *Chicago Tribune*, then the *Sun Times* while he smoked his way through a series of packed bowls.

Mom was remarried by the time I was three. Two of my stepfather's four brothers went to Nam. One worked supply and came back with tens of thousands in stereo equipment and art. The other never spoke a word I ever heard about Vietnam, and drank hard but mellow, a Budweiser man. He married a woman who left him, so he moved back with his parents until the age forty-five. His parents, my brother's and sister's grandparents, were first generation Poles. Like all grandparents of my generation, they lived through the Depression and met during World War II, which grandfather fought in Europe.

Mom's father, stepfather really, was also a Pole. A merchant marine in the Pacific during the war, he got rich trading ivory and gems on the black market. My mother still talks about the small velvet bags all over her house while growing up, the bags stuffed with uncut rubies, diamonds, sapphires, and emeralds, her home decorated with nineteenth-century carved ivory from China and India. Mom's mom divorced him long before I was born when his angry drunken stories of a wife and kids in Japan turned out to be true. In later years he came by each Thanksgiving. He always brought small gifts he'd purchased on Maxwell Street in Chicago, a warren of stalls and tables covering several square blocks, offering the wares of vendors and thieves alike—anything you can imagine, but mostly garage-sale and trailer park trash, rusted tools, and stolen stereos and TVs. One year he brought us these weird small radios (mine was blue) the size and shape of a halved grapefruit. The flat side was a mirror, the top of the dome was holed for the speaker. Sunday mornings I used it to listen to Kasey Kasem's Top Forty Countdown. The same year he gave my mother several boxes of Ramen noodles, the kind you buy five for a dollar and most people eat only in college. Another year several huge blocks of government cheese he'd bought from somebody on welfare, selling it for drinking money. He wore a patch over one eye and would sit and tell WWII stories for hours: sailing all through the Pacific delivering troops and supplies in the Philippines in 1941, India in '42, Australia in '43, then as part of Halsey's main naval supply line system throughout the rest of the war. He spoke about dogfights, submarine patrols, depth charges and torpedoed ships, oil slicks, floating bodies and debris, of kamikazes, burning ship decks, roasted and torn shipmates. He talked about traveling in and around Japan in the '50s, Indochina and Southeast

Asia in the '60s and '70s (family legend has it that he worked as a mercenary after the war), of returning to Chicago to retire to a neighborhood unlike the one he'd left: no more Pollacks, full of spics and niggers. Not that he didn't stay. When he died in the 1990s he still owned several properties in some of Chicago's roughest neighborhoods; kept several pistols around the house along with several hundred thousand dollars in cash. He and my mother were close, and while alive his property and bank accounts had all been in her name, but he left everything to his Filipino mail order bride forty years younger than him. He'd married her a few years before dying and she emptied his bank accounts and safe deposit boxes immediately after he died, before my mother thought to ask about them.

I never planned to join the military, hated the military actually. Mom had raised me to hate the military. In third grade I attended my first parade, a small suburban affair. I paid little attention to the marching bands or floats, though I liked the majorettes and dove for the Tootsie Rolls and hard candy thrown by the waving floatriders. It wasn't until a group of WWII veterans marched by with their color guard and VFW banner, and after them a float holding some sort of large missile with radiation warning signs prominently displayed on its carrier, and my teacher made all the students stand with hand over heart as they passed, and I looked around to see lots of other people standing the same way while here and there a veteran in the crowd was saluting, that I realized not everyone hated the military. (Similarly, at the age of seven, while I knew not all people were Christian, I was amazed to discover not every Christian was Catholic.) In 1987 my best friend, Ed, who I'd known since third grade and who was a year older than me, told me one day that he'd joined the navy. Don't do it, I said. Let's call your recruiter right now and get you out of it. What're you thinking? I said.

Several months later, the summer before my senior year of high school, I got arrested tripping on acid with a group of friends on the roof of the North Riverside cinema where we all worked. Before closing, I taped down the spring lock on a fire exit in the rear auditorium so we could get back in to get to the roof and watch a meteor shower. Milky Way visible, trailing comets streaked across the speckled sky, bright like rockets. The LSD and the beauty of the burning lights actually caused me to gasp in pleasure, and for an hour or more I laid on the roof watching the show, talking to no one, and trying to catch my breath. One of the others got jittery and loud and had to be talked down. A janitor heard the voices from above and apparently figured out we weren't God and a choir of angels because he called the cops, who called the fire department, who brought a ladder truck to bring us off the roof so the cops could cuff us and stuff us into squad cars.

In the car, arms twisted behind me and soaked from lying in a deep puddle from the air conditioning unit on the roof while waiting my turn to go down the ladder, Bowie's "Major Tom" was playing on the radio. Damn cop must have known we were flying cause he turned the volume up and I kept getting swept along with the music as I unsuccessfully fought the effects of the LSD. At the station, locked up and waiting for my parents to get me out, I remembered the army recruiter who'd called me a few days before, who I told there was no chance I'd even consider joining. Two weeks later, I shook his hand when he came to pick me up to go get my physical at the recruitment center. I signed on for two years as a multiple launch rocket system crewman the next day. After all we fought for, Mom said the day I came home, told her that I'd tricked her into signing a form allowing me to join this man's army three months after turning seventeen. It was 1988, the final years of the Cold War.

When I arrived at Fort Bragg, in North Carolina, in December 1989, I was eighteen and fresh out of AIT—advanced training that made me a rocketman. Three weeks after getting picked up at the Replacement Depo—that Limbo between training and permanent party where you wander a shade, waiting for some corporal to come ferry you across the river to your first real posting as a soldier—I experienced my first alert, locked down and preparing for Panama. My parents and younger brothers and sister drove down in a freak Carolina blizzard so we could spend Christmas in a hotel room, ten minutes from base in case word came to deploy to the Canal Zone. While waiting for their hours-late arrival I played football in the snow with the rest of the fellows in front of the barracks, bodies falling hard into the soft cold.

Turned out the Army didn't need the most powerful non-nuclear artillery system in any army in the world to depose Noriega, so we stood down before the snow finished melting. In the end only the light infantry went, taking along their Sheridan tankers and cannon-cocker artillerymen. We all breathed a sigh of relief, having literally dodged the bullet (clichés don't seem so hackneyed when what happens is real). Still, not a few of us looked in envy at the yard darts' combat-patches from the moment they returned until August, when the bullet finally caught up to us.

When it happened nobody expected it, like the random shot that comes through your living room window killing you for no good reason, nor any bad one either. Willie Rivera and I were visiting my family for the weekend, as I often did with friends who wanted to get away from the barracks for a few days. Dinnertime and everyone's eating barbecue chicken, roasted potatoes and corn, grilled onions and green peppers and tomatoes on bamboo skewers. On the news a developing story in the Middle East: Iraqi forces rolled into Kuwait, sweeping aside what little resistance the Kuwaitis could muster. Scenes of victorious troops on tanks and in trucks firing AKs into the air while they speed through the streets of Kuwait City; the Emir and his family in flight, some few of them unable to escape to Saudi Arabia; weeping black clad women, white robed men and boys throwing rocks or running or firing assorted weapons at Iraqi troops; dead people in the streets; looting and flames.

My mother looked at me a long time without speaking. Willie and I barely paid any attention at all. Do you think? she asked. No, we both said, not a chance. Would have to be something big, I said, a real war. Next day was Sunday and time to head back, two and a half hours on the road to Bragg. The entire way, convoy after convoy of military trucks, Willie and I wondering, what the fuck? Must be National Guard summer training, I said. Yeah, Nasty Girls going to summer camp.

Monday came and went with no worries. Tuesday morning, four AM and everyone asleep. Davies, unlucky on twenty-four hour CQ duty, pounding on everyone's doors to call formation in thirty minutes. And there it was. Fort Bragg is the Home of the Airborne, headquarters of the 18th

Airborne Corps and the legendary 82nd Airborne Division. Together and with a few other scattered units and divisions — the 24th Mechanized Infantry with their Abrams tanks and Bradley fighting vehicles; the Screaming Eagles of the 101st Air Assault with their Apaches, Blackhawks, and Chinooks — they made up America's Rapid Deployment Force. The Marines say they're first to fight, small and mobile and ready to go, meaner than mean and nastier than your old man'll ever be, but the RDF is as fast as it gets. The 18th Airborne is so named because it's their mission to be ready to deploy anywhere in the world within 18 hours, not just a few battalions or a brigade like the jarheads, but the entire corps.

My unit was one of a kind, the only airborne rocket artillery in the world, ready to go at a moment's notice. As a result we lived a life full of generals' inspections and training cycles. Each of our three rocket batteries, each with its nine launchers, spent one month on hot status and two months off. Being hot meant you signed out every time you left the barracks and you never went somewhere you couldn't be called, and you never went anyplace that took longer than thirty minutes to return from — and this during peacetime when no one was going anywhere to kill anyone. True to our Corps commander's word, eighteen hours after Davies woke everyone up, we were ready to leave, but instead went into a holding pattern. Ready to deploy and able to deploy mean different things. No one had seen this one coming and we didn't have the planes nor the ships to get all our troops where they were going, and so our first launchers arrived in Saudi ten days after the invasion. It would take sixteen more for the entire battalion to arrive.

We spent our first week waiting to leave locked down in the barracks, doors chained at night and phone access forbidden. Once things calmed some, they removed the chains and allowed us to make phone calls. I'd gone on leave for one month in June and July back to Chi. While there, I ended a longtime relationship with a beautiful dancer named Michele who I'd dated for two years in high school and then for my first year in the army. One day, same as I'd done in high school for countless hours, I sat at my friend Steve's house, in his basement not paying attention while he talked to someone on the phone. Of a sudden he threw the phone at me. On the other end was another girl I'd dated a few years back, Karen, my great unrequited high school love, a girl who should've been one of my great high school heartbreaks, whose name I wrote on my white leather Nikes alongside Cure lyrics, in pitch-perfect '80s postpunk, shoegazer style. She was from Berwyn, but had moved to Brookfield and lived across the street from where Steve and I went to school. Karen was a year older than us and in college like most other people I knew and while I hadn't kept in touch with her, she and

Steve had remained friends. We made plans to see each other before my leave was up, but she ditched me to hang out with her guitar teacher, a hair band wannabe, and I didn't really care. After they eased up the lockdown, along with everyone else, I began calling people. Among the first was Michele, but she had her own things going on and didn't want to hear about my fear. One night, sometime after midnight, I called Karen. We talked for hours that night and every night til I left. I told her I didn't know when we would leave and wasn't allowed to say even if I did, but the first night I don't call, you'll know I'm gone.

When that day arrived, while my battery waited in red dirt and late-summer Carolina humidity at Green Ramp, sitting around or sleeping or reading or staring across the tarmac at the civilian passenger jets that would take us, I asked our chaplain to bless my rosary, a lovely fragile thing of silver with polished black beads that Michele had bought for me years before. A Protestant, he refused my Popery—a shitty thing to do to a soldier preparing to leave for war, I thought then and still do. Even worse, when we got the call to board, while gathering my ruck and duffels, I noticed small bits of light and darkness dropping from me to the ground. My unblessed rosary had snapped, shattered really, and though I gathered what I could find and hurriedly stuffed the remains in my uniform's breast pocket, I knew my luck was up.

EQUIPMENT

While my wife at my side lies slumbering, and the wars are over long,

And my head on the pillow rests at home, and the vacant midnight passes,

And through the stillness, through the dark, I hear, just hear

— Walt Whitman

Dawn breaks open like a wound that bleeds afresh.

— Wilfred Owen

The Five Things
Fire—Direction—Control

The discipline to shoot a man
lies not in the trigger finger,
a quick jerk more reflex
than act of will, but in the eyes
that must be taught not to see
a man holed by bullets fall
spitting blood-soaked thrashing dead.
You must practice this often.

The discipline is a needle, the act
of entering grid coordinates & azimuths,
flipping the red safety cover up
to press fire is the thread
that follows. We practiced
constantly, but when rockets launch
the world bursts with flame & poisoned
smoke, everything roars, shudders
so deep we sometimes tore
at the launcher doors to escape

our own strength. In training,
discipline means cohesion,
unity of men against fire;
but in the shit, with louvers slanted shut,
command hatch buttoned down,
the SPLL dressed to kill, it's every man
in his own helmet, communication
drowned & eccentric, concentric effort
lost, reduced to holding on
to a voice inside saying wait.

The firing done, rockets gone,
the crew sweat-soaked, training
takes hold & it's time to reload,
to release the death that takes
place without & within; to see past
soldiers dying fast or long before your eyes
or somewhere beyond rifle sights
takes patience & time. In basic

& advanced individual training,
sergeants drilled into us
ten thousand things that remain
even today in muscle memory:
ceremonies of measured movement
rituals required to fire a rifle
barked commands that make
the launch of rocket bomblets
& metal frenzy a matter
of practiced mechanical acts.

The aim of training is unity
of purpose, the ability to see
the complete picture through the mist
of a thousand shattered pieces.
With discipline, the moment
impossible to unremember
becomes the one thing you will not see.
You must study all ways.

Jody Call

We knew each other by numbers & letters—
not just the name rank serial number
every GI recites in the movies:
tree-two-tree, six-fife, eight-zero-four-niner,
sir.

11 Bravo. Dogface. Grunt. Infantry. Leg.
At Ft. Bragg, home of the 82nd, home
of the 18th ABC, home of the *airborne*,
we called them wind-dummies, yard-darts.

C-130 rolling down the strip,
airborne-ranger on a one-way trip.

13 Bravo. Cannon-cocker. Gun-bunnies
who could drop a 155mm shell
in a 50 gal. oil drum at 5, 10, 20 miles.
Dumb as grunts, bigger guns.

Artillery: Oh hail, oh hail artillery.
good for you,   Artillery's the life for me.
good for me.

I don't know what they called
us 13-Mikes when we weren't looking. MLRS—
Multiple Launch Rocket System. Klik Killers.
Grid Square Removal Service. When we fired
on Jalibah AB, rockets spread a stormcloud
of submunitions. Target saturated.
Rain-makers danced, reloaded:

Fire mission:	fire mission;
fire mission:	fire mission.
Ready?	Ready.
Fire:	Boom.
Fire:	Boom, boom.

Sometimes I forget. The numbers, the names,
the cadence, rhythm & language pile up.
Basic training: B Btry 1/31st FATC: AIT: D Btry
1/31st FATC: Permanent Party: B Btry 3/27th FAR.
PVT, PV2, PFC, SP4/CPL, SGT,
SSG, SFC—they pile up.

The nicknames, last names, sometimes
first names: Bull—5'10", 230 lbs, a steer from Texas;
Cat, last name Stevens;
Binky from Bingle, Sully from Sullivan;
Frank Niederkorn brought Unicorn with him.
It's all

stand up, hook up, shuffle to the door.
Jump right out on the count of 4.
1,2,3,4.

We filled out wills before deploying to the KSA.
BC gave Top (who gave the platoon sgt,
who gave us) casualty cards. We placed
them in our left breast pockets, below our name tags
or US Army tags—it's all the same.

If I die in a combat zone,
box me up, ship me home.
Pin my medals to my chest.
Tell my mother I did my best.

Many go.

By the third day, we were ready
to go AWOL,
to go to war,
crazy with waiting,
ready

to get on the damn plane
& fight to end the waiting.
We weren't ready

to go back to our barracks,
full of people after another attempted
flight to Saudi, but empty, locked,
& echoes in dull hallways with floors
that hadn't been polished for a week.

Chains on the doors & steel
bands around our lives,
our wall-lockers were closed to us.

Sitting at Green Ramp,

waiting to go & sweating dark circles,

Birdie pulled out a deck

& we played spades—our uniforms

white with salt,

soiled with the dirt we rolled in.

Back at the barracks, Luna washed

his one uniform,

cleaned his rifle,

made another last phone call.

Departure

The postcard enclosed shows
a plane set against impossible blue.
We flew from Green Ramp
toward a kingdom empty
of places I can name,
so I fill these hours riding
through night-black skies
calling to you over & again:
Karen, Karen…

Around me, armed men sit harmless,
M-16s shotgunned, bolts removed;
some stir restless with dreams,
others stare or write first letters home,
& no one speaks except in sleep.

Hard to name the ways I changed
when charging with a bayonet
in training, firing at silhouettes
on the rifle range, confirming kills
while playing soldiers' games.

Harder to know what to expect
after sand & sun get in my head.
Will trees be green with spring?
Or the ground ankle deep
in red & brown, thick with leaves?
And what will summer mean
after wintering there?

Let me get it straight. Days
that turned to weeks spent waiting
to leave created a need for flight
to deserted spaces. A desire
to fight rose up in me, a dread
anticipation poured like sweat
when I boarded the jet
on this postcard I send to you.

In the picture, the plane floats
on a field of cerulean blue,
but I'm taking a jet-black ride,
& can't say whether my flight
is toward or away from you.

Deserted Odyssey
Sentry

On a dune, at night, the red warning light

atop an oil-drill blinks every eight seconds

as I stand thinking about the cool dark air.

Against my boot, a scarab beetle scrapes.

Shining my green-lensed light downward,

I beam the beetle's sandprints,

imagine life pressed into blown desert.

This day is lost in 120 degrees of heat,

the bounce of my HEMTT, Jimmy O rolling

down a hill, knee to chest, rifle flung away & full of sand,

kevlar helmet bounced off his head,

waiting for him at the bottom.

This day is lost in the quiet toil of the scarab.

2

The sun in morning sky burns my shoulders.

Just these last moments have left me sweating.

I wrap a water-bottle in green cloth, soak it to cool.

As it heats, evaporates, I watch.

3

On this hill called Listening Post, I guard

soldiers blinking in sunlight, sheltering in hot shade.

Carlito lent me his goggles before he left,

his duty over. I covered my mouth & nose

with cloth as Sgt Eaddy brought me out.

But my hands are unprotected

from sand & stone. They bleed

as I grasp a burning rock, jagged like coral

or broken lava, throw it behind the bunker

into a pacing space kicked free of stones.

The windslashing sand piles against around over—

scars the ground. I throw another one.

In the distance, a Saudi's bedouin tent flaps green-striped

& white, but here sand strips flesh from bone.

Wind blows over red-striped rock.

Blood dries, sand scours, erases me from stone.

4

If I stand watch, ensuring no one enters

the battery's perimeter, will exhaustion bring

Birdie's dwarf, eager to claw its way from his nightmares,

up the hill to wrestle away my weapon?

Or will my shoulders bunch as I heave larger stones,
filling the empty, emptying the full?
My M-16 lies covered in the foxhole behind me.
Blood slips down my index finger, drips
to the ground. Sand hazes over the dune,
browning the air, & the whip-roar of wind
recalls dead-dry seas.

5

Crisp, chilled—autumn
& a startled blackbird
wings from a branch.

Awake to aching hands,
sun in my eyes.
The wind died while I slept.

The humvee grinds
up the hill
with my replacement.

Balanced in air, between.

The grenade flips peaks rolls
returns to my hand as I learn the weight
tossing it back & forth, hands held
18in. apart. The grenade rises green
to eye level, transcribes its half-circle,
does not hit ground. I pull another
from my LBE. Carlos glances up from the 60,
checks his flak jacket.

                    In basic, Drill Sgt Blinkhorn
grabbed my helmeted head when I lobbed
a practice grenade over the bags—body rolls
to right side, hand at thigh holds
grenade, arm sweeps over head, release.
"Wrong, John Wayne," said Blink,
marked my kevlar: L. Ten minutes
to rub away the lob, erase the idiot chalk.

I never John Wayned another—not when I qualified
expert, low-crawled to a bunker, cooked off a dud
that stitched aluminum welts across my forearm.

Not when I lifted myself to pitch another dud
at a silhouette thirty meters distant.
When I handled two live grenades,
turning one, then the other in my hands
as Drill Sgt told me to remove first the clip,
then the pin, to throw.

                              "Every grenade has three
safeties," I repeat to Carlos. "The fuse is timed —
eight safe seconds after the spoon releases."
Carlos checks his helmet strap.

                              Three now:
two left, one right, toss left to right,
repeat, flow. Chance the air, control each hand.
Carlos hunches over the M-60. His eyes scan
horizon, hand rests on barrel,
barrel swings easy from stake to stake, fills
its kill-zone. The grenades slip through air.

Four Camels in a Line

One-five-

five guys

said their

cannons

blew those

camels

to shit

called it

target

practice

called them

raghead

mother

fuckers.

The road trips drove me sane.

Sgt Lambert in the passenger seat,
I turned up the volume
on my portable radio, louder
than the military radio it sat upon;
I led the convoy with Violent Femmes,
not giving one fuck,
gas-foot to the floor, forcing
my vehicle to its maximum 62mph.
Sgt Lambert asked, "how fast we going?"
& I said, "45, Jack."
And that was freedom.

We fought 1st & 3rd Ammo
for the privilege of 10 hours
on the road to Dammam; we fought
for a break from work details
& chess: sitting on a cot, listening
to the Doors, Tucker sang,
"summer's almost gone,"
captured Jimmy O's knight
& Randy laughed, "idiot."

Riding down sand-blown stretches

of road, we drove away from writing

letters we couldn't end,

our platoon sergeant,

& the unearned sweat

of 130 degree afternoons. I tried

to drive back home, but ended up

in the rush-hour of a Saudi city,

anonymous & crowded

with the Toyota pick-ups

it seemed all Saudi men own.

On the road, we moved,

went somewhere; we didn't sit

studying maps, practicing chemical warfare,

or dodging imaginary bullets

as we low-crawled & ate sand—

artillerymen using infantry tactics

in battles still months ahead

or years behind.

Volleys

The 'dozers leveled desert dirt to beach.
Platoons made up most teams, but we were spring
bucks, a big bunch of two-year guys: the Young
Guns. We pulled boots, stripped shirts to play, & each
of us could block, or set & spike, then reach
for sky & do it all again, pretending
nothing mattered more than to jump or lunge;
to roll in sand; defeats & winning streaks.
We owned the net, talked shit & laughed out loud
at every miss or misplaced hit in games
with lifers. Willie liked to say, this house
belongs to us, each time we took the pit.
Young Guns strutting stuff, we faced every day
with muscular threats, bare chests & sweat.

Deserted Odyssey
Ammo Dump

There's ants they said,

the other crews

from earlier shifts.

I killed them

to stay awake.

Grover (who would

smear an Iraqi's face

with his rifle butt

after a firefight

not yet fought) slept.

> *A hundred meters down*
> *the line, another bunker*
> *& another a hundred more...*

No sleep for me
as the fat black things
crawled walls & ceiling,
swarmed our M-60's nest
& a box of flares.
Spotting them everywhere,
even climbing Grover's arms
& dream-smooth face
(before he shattered
a surrender's kneecap
& shot a third man
still armed) as he slept.

Out the dark, a shot
& a feral dog's yelp turns
howl as the pack turns.

The ants kept coming
though I kicked every hill
to dust, ranged search-circles
round my post. Piles of ammo
rose dark like city blocks
behind my search & destroy.
Grover (months away
from standing flushed
near a flipped truck
huffing air from smoke
beating a dead man
while I did nothing) slept.

> Still, a reassurance
> on the landline: no fighting yet.
> This isn't war.

About Face

Back at Bragg, Jack Lambert, a big Black
who took no shit, kicked us smart-assed
kids in 2nd into the best ammo squad going.
Sgt Lambert drove us to pull preventative
maintenance, check & service our HEMTTs,
square away the connex, sweep the line, do it
all now then disappear—drop out of sight & mind,
working smarter, playing harder.
Jack watched our backs & we had his,
& no section arrived at its drop spots
to resupply rockets faster than Lambert's.
In Saudi, Tebbe drove the chief
the rest of us followed in convoy,
but when T's truck slipped off the road,
flipped over with Sgt Lambert & him in it,
in blackout drive during midnight maneuvers
to beat the heat to which we weren't yet acclimatized,
Jack strode back to my HEMTT, eyes wide,
pulled my AD out the passenger seat,
& thereafter the chief rode with me.
It stayed that way. In convoy, headed to Dammam
or Dhahran to track down the battery's mail

or gather rations, either Jack drove or me,

each keeping the other awake on sleepless stretches,

driver watching the road ahead, AD checking the rear

mirror to see where we'd been, ensuring the rest

kept close order: Tebbe & Tucker, Randy & Charlie,

Jimmy O & Billy Holman, Birdie & Smitty

bringing up the rear—Jack's Pack, brothers armed.

Far from Ft Bragg, we lived close in tents

while everyone in squad platoon battery & battalion

waited for the deadline to arrive.

But stateside days marched to a different cadence.

A letter from Texas changed Sgt Lambert's step

to the rear to fetch his son from an ex

& never return to us, his boys. Something

about the night before the morning he left

remains: the way he stayed awake

as the sands of his desert days slipped away;

his breathing heavy & measured as a command

decision while he sat across from me trying to sleep;

the strike of a match as he lit a candle to pack, asked,

you up? Sgt Lambert's last hours in Saudi,

a torture of indecision: his son? his section?

Something about the marches we're forced to endure,

the routes we take stay with us, dark choices that make us.

Filling Space

After two days digging in a circle
2ft deep, after sun-drenched hours
filling bags with sand, I lined
the hole with canvas, stacked
sand-filled sacks three high,
then set the pole & staked
the hex. I lived in the only two-man
hootch not housing officers
after I convinced Sgt Lambert with bribes
of privacy after months of group solidarity
& crowded lives in Saudi barracks
& red- or green-striped tents.
I set up my cot & Sgt Lambert's,
stowed our gear, sat alone to watch
the day's last light disappear.
Lambert came back later to pack,
talked all night of the end
of his Saudi stay, left the next day.

II

After Jack packed off to Texas,

left the hex to me, I asked

Randy C to take the vacated space

because he rode with me at Bragg

& for a time in country.

Carrender kept a tin of hard candy

he cracked & ground between his teeth,

said he liked the sound of rocks

in his mouth, wanted to shatter glass

with his jaws; Randy spoke a drawl

he called talking blue-grass,

never backed down from a brawl.

III

During off-duty hours, my hootch
became the place the boys gathered
to talk about our lives before,
to trade letters from home, or compare
notes on what it might mean to fight
the war we stared at every day
while training, & nights at my hootch:
when we listened to Marvin sing
what's happenin, brother? as we smoked
beedies until so numb with opium
the question didn't mean anything;
or the times Morrison stagger-sang
through Greek galleries while we drank
the brew fermented in five-gallon
water-cans until the end came.

IV

Everyone had a batch of homegrown going:
a mash of yeast & sugar—whatever
grape or apple juice, dehydrated pears
or peaches we could throw together—
& since the desert gave up little fruit,
we made do with dates & raisins—& sometimes
added fresh oranges stolen from the mess.
The brew produced what Jimmy O swore
was the foulest Smoky Mountain dew
he'd ever had the pleasure of waking from
half-blind & hungover. One night,
Sgt Russell dropped in with *Zeppelin II*
& a jug of the sweetest apple-jack-hooch
I ever drank. Page played & Plant sang
about a livin lovin maid who blurred
with Russell's slurred scenes of Oklahoma
& buffalo roaming Ft Sill artillery ranges;
I fell into a drunken daze, fell into dreams
as he continued to drink & talk
about his wife—who I couldn't make out
whether he loved or hated.

V

More than anyone else, Bill Holman
& Carlos sat out the hours at my hootch.
Kicked to the curb by a girl
he couldn't decide to marry,
Bill joined up with a BS in Poli. Sci.
& one third of a law degree,
in a panic of failed intentions
in Little Rock; he liked Cat Stevens
& Lennon, told stories of spring breaks
with college buddies—like the time
in Tijuana when a long-legged Mexicana
robbed him of two hundred dollars,
& he got arrested but bought a way out
with the rest of his money.
Carlos, from Passaic, called
the old neighborhood Nowhere,
New Jersey, let me know
the place was no Paterson,
barely worth a line or two
of poetry. Carlos had a baby girl
with large black eyes, a dark-curled
beauty whose mother
he couldn't get off his mind.

VI

We filled 3 months with days there,

but at the edge of the new year, we prepared

to trade that space for a border landscape.

Carlos came to me with a bayonet

& a promise we sealed in the quiet

burn of slicing open our palms:

we clasped hands & bled each into the other.

In a steady rain during the next 2 days,

the battery broke camp. I yanked stakes,

dropped the pole; Randy dumped sand

from bags we still needed to cover

the coming weeks lived underground

in bunkers; Bull Hendrix helped fold

the soaked hex & pull the canvas floor

to expose a job well-dug—a hole

like all the others we lost days digging,

ditches we filled with our lives.

The Five Things
A View of the Road

The way was crowded today

with miles of GIs driving

through Al Qayṣūmah

past Hafar al Bāṭin

following the Trans-Arabian Pipeline

straining the straightway

in deuces and 5-tons loaded with MREs,

extra MOPP gear, med supplies,

pallets & cases & crates

of shells bullets grenades;

HEMTT tankers full of fuel

or water, wreckers & rocket

carriers with cranes. The low-boys

hauled a parade of slow tracks

& tanks, outpacing M-1As

racing Bradleys fast-flanking

M109 self-propelleds, 577s,

rocket-launching SPLLs,

APCs riding the roadsides. And all

so long along the way, & everything

altogether-happening-at-once

so that we drove like flocks of birds

or Bedouins with camel herds—feeling

without seeing the distance between

one truck & the next. Learning to steer

during even those free-float

moments when spring-loaded

shocks threw my HEMTT

aloft, & everything not bolted

jounced & jolted, the bruising

& bouncing at times recalling me

or breaking my hold of the road,

& my driving mind found

a fragile focus that made the way

concrete, the procession

of moving troops undefined.

But when the 101st overflew

us all, their rotors chopped thin

winter air & Birdie called out

over the radio to say every

fucking helo in this man's army

was pounding a battle cry

into his head. The Iroquois

OH-58Ds UH-60 Black Hawks

Apaches Hueys Chinooks

added their swarm to our mob

as everyone poured forward

along & above the road
until the whole world seemed
kicked-up, full of dust.

Charles L. Stevens, PFC
Cat's Eyes

1

Sand

windblown

sky

His

empty

eye

2

Damn sky is everywhere,

Cat says, & there it is:

thin blue paling at the edges,

fast streaks & high shreds

& low clouds in clutches,

heavy-canvas tear of a turbo-prop

C-130 grinding overhead:

everywhere bigger than God.

3

His sky

is

white hills

&

blue valleys

4

Islamic reflections of prayer towers
& shacks seen road-blurred

through HEMTT windows, the Trans-Arabian
pipe along Tapline Road, poles & electric wires

that divide sky like tire tracks in sand
stretching back to places left behind,

& Cat dreamdrifts home
to a recollection of flotsam

& white trash washed onto Gulf
of Mexico shores like a stain

halfway between Houston
& Corpus Christi—an endpoint

with a main road that turns to sand-
path under a winter-white sky.

5

In a shamal, sand joins sky

Their common hue discounts
Differences of distance

World becomes desert

6

Sometimes Cat remembers
a time when left or right

behind meant more than coming
or going, a turning toward

that takes away, a failure to stay.
He asks again, are these roads

embedded in skin or laid across
the surface of our lives? Tattooed?

Worn like uniforms? Am I here
or on the way to where I've been?

7

dying day marks the edge
of life's horizon: sun
the size of light at the end
of a rifle barrel: size
of a headshot at three
hundred meters: sky
crusted with blood black
& burnt orange: clouds
soaked like insufficient
wound dressings: night

Deserted Odyssey
Borders

Stand-to & I can barely stand
my frozen foxhole in a February
desert. Around us spring grasses
sparse the stretches, but here
winter stormwaters gather
in a run-off basin before baking
with sand into clay
into which we dug for days
our bunkers & battle nests
for stand-to, & I can
barely stand it here anymore.

The day we arrived, I looked
at the cracked crust, like a dried
lake bottom, to see three days
of pick-axe & shovel,
the adobe-hard ground crumbling
to flakes & dust to choke us.
Tebbe quit digging at 5ft under,
said he refused to sleep
6ft deep—in slit bunkers
already too like graves for me.

Each dawn we rise to stand hunched
& holding rifles under poncho liners,
watching the line in all directions.
But even a thousand meters out
gets me nowhere, so I stare harder—
past the dunes that ring us,
over the hills & farther
than the easy reach of our .50 cal:
beyond mirages & rumors of chemical
ambush, pre-sighted artillery
strike points: to a sand berm
& border crossing.

Please God
Don't let
Me die
Here there's
Too much
Left to
See please

Me & Birdie Down by the Mission

No schoolyard, though we
shouted & laughed some, rifles hung by,
helmets tossed aside as if
the overdrive roar of rockets
downloading wasn't drowned
by the overhead rumble of B-52s
seeking the same Iraqi DivArty
as the 13-Foxes
who were our eyes.

> Lessons came later
> in the shape of burned-out
> rocket engines amid a junkyard
> of still-smoking remains.

Knowing the nature of first
come, first served,
we worked the crane
harder, pulled six-packs
faster, slung tons by touch
in pitch night, so
we could arrive before
our carpet-bombing cousins
who we raced—

to a smashed convoy,
holed to pieces like everything
I knew, while our passing column snaked long
& we tried to keep together.

The Five Things
Floodwaters

The weather means the season
changed to rain, but Ṣahrā' al Hijārah remains
brown beneath sprouting green;
the patches barely mask
the desert's lack, & no grass
grows on our bunkers.
Three weeks since digging in
after pulling out—the weather piling on
as we loaded HEMTTs & SPLLs with gear:
arms lifted in a downpour, hope sodden
just one day into the year.
Now two months gone
& two base camps behind,
like winter growth wind-slashed
by spring rains, we stay close to the ground,
maintain our vigil through below-frozen nights,
blink away the sweat of border-days.
Even when I leave each morning
to check my HEMTT, its rockets,
run the engine to warm me,
or drift to Bill & Bull's bunker
to talk of home or going over,

or deliver maps bearing north to the Euphrates
& on to the next hole with news of direction,
it's always the weather we return to.
Like last night when Tebbe said,
I'm so blind, Achmed could walk
right up, look me in the eye
& I wouldn't see—then took
my pawn with his rook. Check, he said
& I did: outside the red glow
of our flashlit circle: all around:
up to see stars rubbed black
before realizing it must be clouds.
In the sabkha we occupy, surrounded
by sanded ridges, a steady rain
means a gathering of spill-waters
run off from dunes through wadis to us
bunked down—or like Tebbe & me:
awake with a machine gun, eyes
to the surface we could see, the pieces
under a heavy lens, the first drops
as our watch stopped. We collected
the game, our rifles & webgear, picked
a dark way to our hole for sleep
while it came down. Slow waters
flowed round the battery. Sand-bagged

berms unable to contain the weather
collapsed in noise & heaves. I broke
surface like a swimmer seeking air; soaked,
I slogged through to higher ground,
& now wait as the morning quarterlight
shows wind rippling across lakescape.

21 Feb 91, Evensong

Karen, your letters arrive like clouds—
after I've stopped waiting—
or while looking for something else—
on guard at 4am, some wisps hanging
in predawn sky—me stupid with grins.

Used to be I counted days—
4 Jan was Birdie's birthday;
he turned 20. Just think:
258 days until your 21st,
259 until my rebirth as a civilian.
Today makes 8 ½ months
before that much-counted-toward time
arrives. One day I'll count away
from numb hours lived in letters
where time isn't as it looks
& places we two knew fade from view.

For months after entering the desert,
we expected to return to the world,
but the talks at Geneva failed,
& even rumors died while we prepared.
Makes me wonder who will forgive
the death we're equipped to deliver.
Can you see me reloading a launcher—
the Guard captain chattering to captors—
about metalstorms at the end of creation
& no command & everyone dead—
my fellows & I excited
by killing & not dying? Later today
we move to position by platoon;
maneuvers begin tomorrow at noon;
we expect to fight soon.

When we first arrived,
after our first field exercise,
forever-open views from the heights
of dunes convinced me
nothing could hide from sight.
But the desert has its own way
of disappearing truth
over the next ridge in wadis

behind man-size clumps

& bumps where strange succulents

put down roots, understandings

you either move toward unknowing

as in the moments before ambush

or secrets scraped from sand

when in a digging state of mind & body.

So because I've dug

with pick & shovel through sun

burned clay all day longing,

knowing tomorrow brings the same,

I can say I love you, knowing

tomorrow's effort awaits & there I'll be.

Haven't written about the B-52 noise

in the background of our days

like the words I imagine saying

as I concentrate on you, always

falling dumb like bombs

when I try to figure where you'll be

when I relive these days with someone

in bed beside me.

It occurred to me that sight happens
in time with time. Things now
I didn't see then. Our lives turn back:
me in a town neither of us recognize,
in love with a different woman's
photo of a worried 2 month-old
in her mother's, not your mother's arms,
thinking of you then & me as I write
these words.
I get in my head when I sit
through sandstorms, poncho liner flapping
in the bunker's entrance,
Tebbe's knees inches from mine,
we two hunched over book, chess,
nothing, or writing letters about life
after war—but seldom a word
about the desert pushing past our efforts
to contain or at least deny. I apologize
for the eyes that make me speak
in definitions; the spirit—
our drill sergeants told us—
of the bayonet is to kill—
or at least get the point across—
a GI joke.

Strange to think someday
you'll never mean more
than you do now. Or rather,
the day may come when letters
from a border bunker are all I retain
of you, the rest given back
so that the living fact of you means
less than the dead who shared the war
with us. Stranger to see the remains
of you and my army days leaving me
like slow shell fragments rising to break
skin built thick over years—not now,
as I write, but years from now.
Just as the night breezes that cool
the end of days can't compensate
for what we pay out in sweated efforts,
Karen, if you leave,
though your letters stay, it isn't the same.

You should have seen the sun rise today.

As we stood to, it rose under
a checkerboard sky full of small clouds,
turning some into red & bleeding dreams,
others into black silhouettes like memories.
I thought of my first weeks here,
a vigil strained waiting for shade clouds
to blot the sun, clouds I watched for
the way I scan your letters for words
like love, words to let me know
what we'll mean after the sun sets
calm in the open, behind dunes,
the world awash in gold orange white,
colors without definition, in red.

The Five Things
Assault Dimensions

The terrain cannot be seen
in rocks that lie abrasive
in layers atop grit & dirt
that makes the land here.
Jags tear our tires
while the clouds we raise
like wind ruin our view,
& the true terrain escapes us.

As when we landed expecting
to depart for sands days after stepping
into a heat-blasted kingdom
of deserts we didn't know or yet
understand until the topography
turned around as summer settled
& sun bled to black winter
nights when we didn't sweat
the terrain though straining to see
in darkout drive, distance dwindled
to three meters between vehicles,
dimensions set in tints of night-vision-green.

The day we passed through
the barrier, Randy & I shook hands
before mounting up; all the dunes hid
in clouds for hours after Colonel
Thrasher broke radio protocol
to welcome us all to Iraq.

In our sector, 2nd Platoon crossed over
the border's berm in a shamal;
we pushed past the wind's last sands
to a black-blue sky above a plain
covered with ragged runs
of rocks forever-in-all-
directions—the terrain
exposed & us in it. Our tracks

flowed fast across the flats.
The Euphrates is where it began.
The land rippled circuitous.
Tributaries dry, with twists.
Alpha became bogged in the loose
sands of comprehensive dunes
while Bravo tore up
their half-mile sides
to enter the River Valley firing

as Charlie scattered under Iraqi
artillery, someone on the radio
begging for a tow as a howitzer
walked rounds down to him.
Launchers in the valley firing fast.
His voice in static drowned & cracked.

We gathered rocket covers for souvenirs.

The trailing rockets
pulled roars through sky.

I smiled along,
releasing the slow-rumbling distance

between collecting memories
& torn Iraqis dying

while munitions rained
down. Ground vehicles men

exploding as shrapnel
ripped sand into air.

The sun sets a different way when you're flying.

Did Steede notice
the color of sky
when the T-72
round blew the turret
off his M-1,
blasted him through concussed
air? Inside, clutching
his legs, the loader
fingered wounds, pulled
sizzling shrapnel & flesh
from holes, not
knowing steel from skin.
Driver screaming & blind.
Sgt Sarringer afire,
melting to gunner's seat.
Did the force
of landing jar Steede's
senses? Could he
see black smoke billow,
hear secondary explosions
of stored ammo, smell
his buddies die?

I never witnessed death before Jalibah.

Coughed from a rifle muzzle,
the bullet—in one direction—
from rifle through body into sand,
the spirit of death in war.

2

In broken & shattered
remnants of a T-72 tank,
the man lies dying,
counts his breaths.

3

In the mirror-brass of a lead bullet,
dunes piled along
the Euphrates River Valley
shifted in the wind.

4

Digging a fighting position in a sandstorm,
I listened to the wind. The dust-devil
blew past & around my goggles.
I couldn't see to fight.

5

We stood heated by blasts of desert air
in a bunker near the battle front
while the thunderstorm churn of rocketfire
chilled our blood again.

6

When we spoke, we spoke of rockets:
Get those six-packs downloaded.
Fire mission. And rockets downrange.
Today, bodies lie buried beneath sand.

7

The Iraqi with no face
only whispered,
but I couldn't speak his language,
so I left him to die.

Richard Gregory, SGT

Rick the prick had the balls to tell Sully,
call me dick, but don't think you'll forget me.
As much as you try,
til the day that you die,
I'll be there, dug into your mem'ry.

Though he may have been right about that,
his ass nearly got fragged in Iraq:
I spoiled his fun
when I showed him my gun,
told him, Sarge, I'll be watchin your back.

G(ilgamesh)-plus 3

When it darkened all day
we sent up rockets

to sow their fields with fire
before we advanced on Ur.

We mounted the ziggurat
after tanks rolled in

& over every camelfucker
left alive to know what hit im.

M-16A1 Combat Assault Rifle

I never fired a gun before
I entered the army, but as I aimed
my M-16, I knew I would kill
the Iraqi soldier. He danced
in a daze, bled from a head wound,
tried to surrender, held no AK-47.
He dodged bullets in my sights & I held
my breath, my trigger:
closed my eyes, breathed in, held,
breathed out, held, opened my eyes
to meet his. There are days

never-changing: moments
stutter & stop, flash & back up
as they happen: days we played
baseball in Ed's backyard
lasted all summer. His granddad
nailed up plywood to keep
us from knocking boards out
the pine fence behind first base.

But plywood didn't stop homeruns
that dented the hood of my mother's
mustard-dust Delta 88, didn't stop
my father's belt nor hide my bruised back.

Fall, we played football weighed down
with shoulderpads & helmet.
One afternoon I called Dave, fucker,
again & again as he hammered
my back, ribs, & helmet until my nose
bled onto my white Cowboys jersey
& my ears rang. I could barely stand,
but never stopped saying fucker.

Sun still glares then as when
I knelt aiming: saying fucker:
waiting to ease back the trigger:
the Iraqi's life on fire
before my eyes & his.

The day before, my platoon
launched sandstormed missions.
I sat in my truck, tied a cloth
over my magazine dock
& ejection port to keep out grit.
Over the radio someone spoke
of T-62 tanks & incoming artillery.

In the wind, M-16 shotgunned
to brush dust that clogs the bolt,
returned winters when I played
guns with Tommy & Walter & we
threw ourselves into snow drifts to see
who could die best. Alone, I watched
rockets fly faster than fast clouds
shooting across a winter sky:
as when long ago covered in snow,
not so cold & wondering why.

II

Because after removing
starlight scope eyes
that bring to view
a green world alight:
burning tanks & trucks
rocket-assisted rounds from 8in guns
25mm tracers like lasers—
star-thick sky blinks
into slow focus as your eyes
open to no desert no war
no it me you anything:
to silence: to sand.

At hand an M-16: even as I,
broken to components: its sights my eyes:
a seeing, seeking thing:
like a definition: but trick knowledge
because though we ran the land
& owned the whole hard night,
the desert was no home—
nights so dark & given to roaming:
waking lost: off the mark.

What changes & how much remains?
In late summer 1983, underneath
the streetlight I talked with Ed
about *Empire Strikes Back*
& Gina LaHockey's 7th-grade breasts.
In boot camp, we sweated
late-August-Oklahoma nights
as we slept with our rifles
while our fathers, our drill sergeants,
taught us how to kill.

What would you do? In a strange time?
A fast land? We cleaned black surfaces.
With air hoses we blew out trigger assemblies.
Our rifles shined at night.
We learned to respect them.

With brass brushes & oil,
we rubbed rust from days
stored away. Always sweating
through a choice or chore—regardless,
a decision to act or stop acting anymore:
but with a rifle in your hand:
aiming hard: breathing slow.

Elegy in Disbelief of His Witness

Because that day:
after the ashes: smoke
in mouth; flakes:
crisp burned to sight;
thought: inhale: body: drive-
induced—rather reduced—
to form the corners
of my eyes which stopped
trust; not even

the T-62 that hazed to view
as from behind a dune
an Apache en pointe
chained the tank to pieces
of sunlit slants
through corpse-stuffed hatch
& awash in the concussion
my HEMTT shook me
to see ever even.

Deserted Odyssey
Quit Paradise

We commenced fire at oh-five-
hundred to kill time til oh-eight,
then ceased fire all day & forever.
The day the war died, I pulled
guard with Tebbe in a shellhole
on the highway to Basra
strung out on no sleep
strung along a road littered
with dead men & burning Soviet BMPs
Frog missile carriers T-62s
troop trucks. And along came
a car with a pickup behind.

2

I wondered where coming
or going while advancing the days before
with walking wounded & surrenders
on all sides shuffling along the way,
miles from there, hours til where
we drove arrived, the wanderers behind.

3

Tebbe turned toward me

when the car halted,

pickup behind, opposite us

on the road, the sky

bright in February

the day the war

died. I knelt in sand

& Tebbe turned to them.

4

Because after the day before,

after we broke with the Guard

& small arms gave way

to rockets, our firing commenced

near a farmshack, our massed launchers,

our many hundreds of rocket trails

& red glare & smoke

round a ragged fence & black clad

woman shielding two children among

us, & in the distance at the front,

our submunitions like thunder

among men not coming home.

5

And a man & a woman
& a boy & a girl, & Tebbe turned again
to me to see what to do
with the pickup behind,
smoking vehicles on both sides
of the road. Everyone starting to move:
Tebbe toward them, waving & raising
his M-16; the man the woman the children;
me covering, rifle to shoulder,
sights at center mass on man woman children,
the hour still early, the war hours over.
Family, pickup, Tebbe, me: in the sun
in February.

for Austin, born today

69

Homo Furens

Strangest thing, a small owl
tame & tucked away
in the hydraulic innards
of an abandoned cannon.
Two guys from 1st
stashed the stealthy bird
in the empty brass
from a 155mm round.
Top found them out,
ordered it set free.
Sgt Gregory dumped it
on the ground. Big-eyed,
the bird only swiveled
its head around, fluffed
its chest feathers, loosed
a screech eerie & alone,
but wouldn't move. Dick
tipped the owl with
his boot to shoo it,
but it only stared,
so Gregory stepped hard,
catching its left wing

which popped from its
shoulder. Again, a screech
as the injured thing
flapped across the sand
before it sat, folded
its one good wing, nestled
down as if settling
into itself. Picked up
by wind, bloody & broken
feathers kicked across desertfloor,
signifying nothing much.

Charles Hendrix, PV2

I

Barrel-chested Charlie
sang like Hank Sr,
planned to marry

a preacher's daughter
from outside Dallas
who raised horses,

could rope a steer,
ride the bull—
a beauty Charlie swore

he didn't deserve:
a high-cheeked cowgirl
with bedroom eyes

& loose red curls
that Charlie loved
to wrap around

his calloused hands
& brush through
with thick fingers.

II

Before enlisting, Bull
taught Sunday school,
was baptized anew,

refused to drink
the devil's brew.
Charlie claimed Christ

drank the blood
of pressed grapes,
but never wine.

After a baptism
with sand & flame
during our war

Bull decided he'd
seen some things
that made him

think about Jesus
after the cross,
needing a drink.

Where's Joe?

She knew her son wouldn't die
in Vietnam; that was his father's war.
Still, Joe races around
the New Bern HS track—
red oval with white lanes—

the number 28 safety-pinned
to his jersey, Nike running spikes
tearing the red rubber & tar;
blood on Joe's shoes & the flesh
of the race clinging to his rats' teeth spikes,

Joe races on, headlong on
through a haze of 5 days' fighting—
in desert camouflage, his LBE
weightless, canteens filled with mouthwash
to soften the sewage, bouncing & bruising.

The fighting over, rifle slung over
Joe's shoulder, barrel pointing down,
"Get Back" scratched in rust, he moves past
his Abrams, past the perimeter wandering,
wondering about minefields

as his comrades form up

to hear echoes of his twice-called,

twice unanswered name,

seeing First Sergeant's mouth

as he sounds off Joe's name.

Battle Royale

Just this: the whole
of us grappled together:
last day in Iraq,
arms locked: we pushed

back lives lived outside
fighting that had failed
to make us the men
we knew as boys.

These desertweeds are me. Please
forgive this, the letter I never
thought to write. In the Sandpile
summer days stay through the year,
though night changes with the season.
But as snow continues to hold in Chicago
& days remain cold without regard,
I send this letter to you, to tell you
the war is over; I return soon.
Those photos you sent, I hung
in my tent, then taped in my HEMTT
when the leadership sent us forward,
then into Iraq. But pulled the pictures
down because your face grew confused
in my head with the dead we moved through,
& your eyes & theirs stared,
though not in the same way;
so though their scorched grins
were not your smile, I pulled down
your image when the difference faded
just the same, as the miles lengthened
to days, as sleep became a dream of driving.

Karen, this is how worlds end:
the desertfloor the sand
the ground on which you & I stand
thrown up, blown rocket engines hurling
ahead of whickering bomblets,
the sunbright sky suddenly heavy
& black with enemy rain until every truck
& unfriendly tank halts holed
as everyone on this incredible march
comes to a shrapnel standstill.
Those who die, die hard.
The lies & threats that brought them here
disappear. Only the weeds remain,
unused & useless. It comes to this: equipment
counted signed for waiting replacement:
them, us, other GIs, our names, all
lives trying to outdrive the night.

Picture yourself returning.

Though I stopped
To count ten
To see when
I stepped down
Stairs to ground
To look around
Over at spectators
Awake & welcoming
For our late-night
Arrival gathered waiting
It's the back
Of some soldier
I don't recall
& my new
Desert issue boots
I remember when

I remember now

That my parents

Drove to Bragg

Later that day

To drive me

Still seeing Iraq

Past new-leaved trees

Miles of pine

I stood uniform

In their home

They colored eggs

For Resurrection Sunday

& that's it

What I did

Or what said

I can't say

The Gulf

After deploying back to the world, I read poems by lazy warriors, self-congratulatory poet-activists, worse than the recently activated reservist—fat sweating loud—I met at the PX in Dammam. Taking images from TV & the rage of Achilles, rather than rifle sights & night-vision goggles, they imagined a desert storm. On CNN, they watched Schwarzkopf, Aziz, Bush, Hussein, ignored Iraqi grunts living in filth & dust like us. (Still, everyone wanted to know a soldier in the banana wars & my father, who hadn't spoken to me in years, sent letters to my desert address.) News-watching Americans saw smart bombs, championed Iraqi conscripts, never considered American economic conscripts. I never heard Dan Rather, monstrous in his gas mask, say, live from somewhere in Saudi Arabia & I never thought of ancient Ilium, but I watched a truck full of Iraqis charge the battery, & many hundreds behind firing AK-47s, taking fire in return. I watched Chuck move from corpse to corpse to check for breath. Outside his bunker, Pvt Hunter prayed on his knees as a Tomahawk cruised over our border position, certain it would hit; & he never once seemed Agamemnon seeking god-sign on plague-ridden winds.

Elegy

1

November 4, 1918, the British

Tommy slips in mud on the east bank
of the Sambre & Oise Canal. He drops his duck-board
at water's edge, listens to the stutter of German machine-guns.

Three nights ago, it started raining. The canal flows
a foot higher around his leg-wraps, washes
stinking trench-clay. Until this moment, the angry guns

the wailing shells, the rattle of rifles have been poetry.
But scanning explosions leaves hard beats & stress;
there is no rhyme in the cough of a machine-gun;

& after, all quiet.

2

The crunch of sand, felt more than heard,
lingers, more a trochee than an iamb.
Death in the desert doesn't change
& I see no difference between the dirt-covered
face of this American, dragged through
the streets of Mogadishu,

or the dusted, wan face of an Iraqi soldier,
flattened from the waist down,
vehicles rolling over him,
leaving a lipstick-red smear in place of his legs.

On television, the senator from Texas says,
"That boy was tortured," stresses boy, tortured.
The dirt-gray GI wears pants & handcuffs.
His face swollen, lips peeled back.

I think about the taste of sand, wish for a rifle,
wonder who left him behind.

3

Out on guard, I shared cookies & water with four camels.

Earlier, a group of Saudi children begged M&Ms.

Randy fired a shot to flush them from his truck.

Fifty meters from me, a herder dragged a dead camel,

around & around, behind his Toyota pick-up,

over a dune & out of sight. I ate my cookies & hummed,

daydreamed of strange canals & slipping under.

Michael Sullivan, PFC

Still, I think of mail call & Ft Sill
& "Drill Sgt, here," when he calls
on the phone. Sully, from Ohio,
always stands in memory:

when he sat in a sleeping lean
against me while we bounced
through dust in a deuce & a half,
his smile demanded he stand.

Even as he crouched in aim
or bent to fill a sandbag after days
of filling sandbags, it's his back
straightening, his face broken

with a grin somewhere between
"Fuck you" & "Yes, sir"
that I remember while he tells
me again how one night his legs gave,

his face twisted with tics until he fell
asleep, exhausted after hours
of rocket-fire-day returned
to new-moon-night, arm stretched

toward his humvee's cut-off switch.
The humvee, motor running.
"Sometimes," Sully says, "I still hear it."
The engine, awake, at night.

William Rivera, PFC

Willie left Chicago's North Side
because he didn't want to die
a gangbanger like Small Change
who got rained on with a .38,
or Hippo—*un Puertoriqueño*
muy gordo—who some Players
ran down & ran over like a rat
in an alley. We all wanted
to grow up, move on, get over, get away
from somewhere or somewhen.
Willie signed on to stay alive.
He raised his right hand
& ended up like the rest
of us at Bragg.

II

Like the rest of us who ran together,
Willie & I were brothers;
but Willie drank hard, could really put it away—
vodka during duty, beer or rum at night;
didn't mind fighting whoever he could find
when drunk blind. One night he came at me

with a swing & a miss, then another,
then strike three. Willie kept at me
until I threw him down, sank my fingers
in his throat, choked my friend
until his head lolled, his eyes bulged red
then rolled white, & two guys pried
me from his neck. Willie almost died,
awoke on an Army psych ward, found himself
walking in a robe & foam slippers
to greet me, trade apologies with his visitor.
He explained Uncle Sam's answer
to AA traded Phases for the Twelve Steps.
He counted them like a cadence call:
One, Two, Three, & lockdown in a looney-bin
makes Four. Willie got out & got drunk
in celebration, but when the battery
deplaned in Saudi—an entire country
drier than any Carolina Baptist county—
Willie apologized for dragging our asses
with his to Phase Five.

III

When the battery marched forth
in support of the 24th Mech Infantry,
our rockets blazed a tank-trail toward Basra

& a battle with the Hammurabi Guards
that didn't happen once the war ended,
a fight that faded like sleep deprivation
hallucinations edging Willie's sight—
phantom jets & smoke trails that reminded
him of the DT haze back on the ward
or smoking PCP back in the day, on the block
with friends who died. But Willie survived
it all: bats & knives on the North Side,
Army bullshit & Iraqi bullets. Willie saw
past the shamals; in blinding sand
without even a hint of the path, he saw
it through, kept it cool. That was Willie:
driving away from one place, ending up
in the shit: from Chi-town streets
to Army drunk-tank to Saudi sand-trap,
& then Iraq: sand-blind
in a storm—but somehow still
slowly heading home.

Ahead, Through a Haze of Smoke

If I live in fear, it isn't because Jay Clemente
waited behind the foundation of a torn-down two-flat.
Not because he tackled me as I walked past,
held me down, grabbed my hair, slammed
my face into the sidewalk. We were nine
& he beat me eight more times
before I turned twelve, but when I tremble,
I don't see Jay's face.

I think sex every time you slide
from under the covers at 7am.
Tired, topless, you walk to the bathroom.
I resent morning abstinence, but you
dodge my cold hands touching your back.
If we wrestle & have sex,
if I hold you down, are you afraid?

Waiting for my father to come home, I cried.
He opened the door to my bedroom; I looked
at his belt—black leather, two inches wide.
He turned off the light, put a hand on my arm
as I lay on the top bunk, sat down below me
on my brother's bed. In the dark
he never spoke though I listened, learned
that fear falls & rises, like breathing.

The first time I saw a crayfish, it grabbed the soft flesh
between my index finger & thumb when I reached for it.
I shook it off. Ed & I learned to hunt them,
shying away from their pincers.
Once, Arlo & I climbed to my roof,
dropped a dozen crayfish off the side
into a Hills Bros coffee can. Mostly we missed,
killed them all.

Behind the barracks, Birdie lit a cigarette.
The dry tobacco crackled as he inhaled.
"Fear smells bitter, sharp—
like sex-sweat—
like hot cinnamon," he said.
"A hunting dog tracks that scent."

hiding from the sergeant of the guard:
step into a hootch: sit on a cot:
arms crossed over my knees:
head in the bend of my arms:
fall awake at the edge of a hole:
a nightmare: cannot see:
will not raise my head to see:
a high-pitched whine pulses & sweats me:
certain that one move will topple me

Not a nightmare:

Willie walked over so I swung down

from the cab of my ammo truck.

He smiled under his kevlar helmet.

Ahead & to our left, another T-62 exploded

as an Apache stitched holes through it,

30mm tracers laser-beaming from chain-gun muzzle

through tank, setting sand afire.

"I didn't sign up for this shit," Willie said.

I will not apologize for the time

Walter Hawkins & I lured Joey Zabinski

onto the roof. Walter threw Joey's coat

onto the garage. I jumped across.

Joey sat down, cried. His hands shook.

I laughed & kicked the coat over the side;

for a moment, it billowed flag-black.

Walter pushed Joey off the roof.

He fell 10 feet; he screamed like someone

falling 10 stories. I didn't think he'd stop.

First Snow, 1997

Watching the sky bleed light, I anticipate
the evening, the next wait
for an AK's bullet that missed
then & can't catch me now.

If she sat beside me, I would ask
her to walk through last night's fall
because tomorrow's sun melts
today's snow to mud. As we are.

Still awake to see the false dawn of pre-dawn.
Tired & waiting for something missed
years ago. With colder days, the snow
should stay. I hope it hits soon.

When the Burning Sleeps

I. We

Scattered by hand-grenade
dreams, I confused your weight
for a dead man's. You pinned me,
focused your breath in my ear. I realized
the roll of your hips spoke desire,
& our fear at my arousal
scattered me again.

Some nights, I scream with the bayoneted
boy who fades from the space
where you sleep.

Worse than the combat-speed
night-fears of wars not even my own;
worse than sitting on the bathroom floor
because the space is so small, so bright
there are no unwatched perimeters;
worse than knowing the way a bayonet slips
into a knee joint is letting go.

II. Like Those Soldiers

All the exits begin to wear,
so that lately I find myself waiting
for the next person to leave,
pieced into the past
the way yesterday-comrades
puzzle more into place, out of life
with each distance-day

until Birdie rumors lost in Blue
Kentucky; Austin drunk
on the phone, then nothing these long years;
España in New Mexico; Vallon who never left
Bragg; Randy, Jimmy O, Lopes,
so many gone—
except the nights they fight through

desert-smoke dreams with me,
together before I shake awake,
sweat replacing my blood or theirs,
all shouts my own, these arms that held
my lost soldiers as they died, reaching
for the woman who cradles me
as she whispers me back to them.

III. Walking Through

More than any other, I remember
the Indiana field jacket I played GI
in, patches sewn jagged from mismatched
branches covering a sweater, gloves, scarf,
while snowflakes swayed, thicked
through Chicago alleys, floating in lamplight
a brilliant moment, spun in backyard
gardens, garbage cans, alley dogs loped
& growled, but me with a warding stick

long gone — green denim
now olive drab, combat-patched
in exchange for jerk-awake nights,
a decision to get out of bed,
walk down the street with a new field
jacket sewn straight, matching
eleven year-old memories
which burn sometimes — intense
on a flaking night.

I've been. It was.

Trying to get
the bodies out
of my head,

sick of wandering
nights, of mindfields.
Afraid to turn

away when torn
legs arms faces
make me whole.

The Five Things
February Forecast

The leadership never imagined
I would hike through hedge country
to sit near power lines
watching wind-heavy high grass
covered in drifts, whited-out green
in clumps like small dunes,
in a snowed field with upturned face
to ask forgiveness from men I helped
to kill. I've burned years reclaiming
months in sand & sun
with men I no longer know;
wasted days with Drill Sergeant
Graves saying, you will
want to understand
the terms which describe
authority, as he pulled
hand to brow, arm angled horizontal
to teach us to salute; retraced
routes through those hundred
hours thousands of times;
always returning to the seconds
required to aim to shoot

with eyes direct, as when saluting
or staring on watch.
The leadership spoke strategy,
but our tactics were rockets
& steel rain blasts of shrapnel
like a tearing blizzard
the air itself a shredder;
we maintained distance,
fought behind a mask
of Bradleys that once lagged
& the Guard hit us: our M2 escort,
turret turning, blazing tracers; Davies
pumping M203 rounds, explosions
like crushing embraces
that continue to wake me
nights my own mask slips.
The leadership waxed expansive
but I try to pare it down, see things
drop away like a wadi cuts
a landmark into desertfloor;
a desert bowl too wide for sight to fill
until full of Bedouin herds
& later, Republican Guards
defined by M-60 sights.
Storms always, on the horizon

a bermline stretching home.

Midway to Iraq at midday:

a long road trip, a black

turbulence not yet in range,

driven toward a moment

I refused to shoot—

& am what I was when not warstuck,

when I wasn't defined by closing

eyes, before I learned to wait.

Darien J. Wiggins, PFC
16 March 1971–10 February 2008

More than twenty years gone

since we went to see COC

in Fayette-Nam, NC,

& you shared a smoke with a roadie,

holding it cupped in your hand to shield

the cherry that brings the sniper's bullet,

like we learned in the desert.

You offered him 100 bucks

to steal the band's banner after the show

& he laughed and took your cash.

And the drummer strode on stage

& the bass & guitar too

& the singer was the roadie.

    And you grinned that grin

    that cracked your face wide open

    turning your happy eyes to slits,

& said, I shoulda known—no wonder

he looked so goddamn familiar.

We stood on edge that night,

on the fringe of yet another pit

of bouncing swinging thrashing punching

swirling screaming drunks and straight edges

and all the angry boys and girls in between

corroding our conformity

as the band churned on, speeding

the spin, feeding the frenzy,

churned on while you & I stood inviolate,

poised at the edge of out or in.

I realized my boots kept sticking

to the floor cause we were standing

in a puddle of blood from the kid

they'd pulled bleeding from every hole in his head

it seemed & I punched your arm

& pointed my fingers like a gun

at the ground and you looked down,

lifted first one foot then the other

like you'd done that day outside Basra

when you'd leaped without looking

from your HEMTT & landed

in the puddle of a dead Iraqi soldier

& you looked up at me & grinned.

      That grin, your face cracked with grief,

      your eyes turned to disgusted slits.

And the war came looking for us & we went

& the war went & we came home

& the months passed & our army days shortened

& then it was time to go. I tracked

you down on some motor pool duty
my last day on base, you still with months
to go. I'm right behind you, you said, grinning.
    That wide grin that cracked your face,
    your happy eyes open to the days ahead,
    like a swimmer ready to leap
    into the lives we were so ready to lead.

for Mason, your son

Deserted Odyssey
A Prayer

I stood last post on line at a wash stop
for heavy vehicles covered with sacred sands
the Saudis refused to let us remove
back to Christian lands. We cleaned

ammo trucks launchers humvees,
a ritual for our equipments' return,
though we neglected to scrub the war's
dirt from our own hands before birds

of pray returned us home. Allah's dust
stayed with us in places rubbed raw
or too deep to dislodge, stuck in pores
like charcoal from chemical suits,

or the burning oil & steel & soursweet
stench of Iraqi meat cooking in death
that plants itself in the nostrils
& back of the throat.

2

Lord, as when sands cease keening
as winds die & the shamal
slows, blows over us no more dust
that forms desertfloor, so let it be with us—
who out of our depths have cried:

Lord, hear my voice.

An Iraqi prisoner captured
after rising up armed from a bunker
in our midst like a miracle we never saw
coming or a devil sick of sin,
looked at me to say:

Tawakalna ala Allah. Tamam.

Let an evening prayer
be said for those not going home,
let it strike You blind
Who lives & reigns forever
& ever in empty skies.

It is finished, let it end

A note on the type

This book is set in Electra, a Linotype composition used by the poet and scholar Jake Adam York, a dear friend who died in 2012 at the too-young age of 40. Electra was developed in 1935 by William Addison Dwiggins, a prominent typeface and book designer best known for his work as an illustrator and commercial artist. Most active between the Great War and World War II, Dwiggins developed Electra to represent the modern, the machine age, to be electric, to read like sparks, like slivers of metal flying off the page. Just like Jake's poems.

www.ingramcontent.com/pod-product-compliance
Lightning Source LLC
LaVergne TN
LVHW091223080426
835509LV00009B/1142